LOVE BEES

LOVE BEES

A FAMILY GUIDE TO HELP KEEP BEES BUZZING

VANESSA
AMARAL-ROGERS

First published in the UK and North America
in 2019 by
Leaping Hare Press
An imprint of The Quarto Group
The Old Brewery, 6 Blundell Street
London N7 9BH, United Kingdom
T (0)20 7700 6700 **F** (0)20 7700 8066
www.QuartoKnows.com

British Library Cataloguing-in-Publication Data
A catalogue record for this book is available from
the British Library

ISBN: 978-1-78240-664-8

This book was conceived, designed and
produced by
Leaping Hare Press
Publisher **Susan Kelly**
Creative Director **Michael Whitehead**
Editorial Director **Tom Kitch**
Art Director **James Lawrence**
Commissioning Editor **Monica Perdoni**
Project Editor **Joanna Bentley**
Designer **Clare Barber**
Illustrator **Sarah Skeate**

Printed in China

10 9 8 7 6 5 4 3 2 1

Contents

Why we love bees

Bees are amazing. We might not notice them all the time, but these fuzzy little creatures buzzing around our flowers are very important.

One out of every three mouthfuls of our food has been produced with the help of bees. It's even more if you include the clover and alfalfa munched by the animals we eat. If there were no bees, our dinners would be very boring indeed. We wouldn't be able to enjoy some of our favourite fruits and vegetables or many of the beautiful flowers we see around us. Of course, without honeybees, there'd be no delicious honey to spread on our toast!

Bees visit flowers to get something from them. They love nectar, a sweet liquid made by flowers to encourage bees to visit and pollinate them.

VIOLET CARPENTER BEE

I'm a busy buzzy bee!

FACT
Not all bees are black and yellow. They can be all kinds of different colours including blue, orange, green and even purple.

Nectar is made by special parts of the plant called nectaries. Bees also eat the pollen made by the flower, as well as collecting it for their young. Some solitary bees mix pollen into balls with a little bit of nectar to leave in the nests for when their eggs hatch, so the newly hatched bees have food.

There are 20,000 different types of bees in the world and they look very different – from a tiny mining bee just 2 mm (¹⁄₁₆ in) in length to a giant leafcutter bee that can be 1.5 cm (⅝ in) long. Bees live all across the world, in every continent apart from icy Antarctica.

POLLINATION

Pollination is the way that flowers make seeds, which can then grow into new plants.

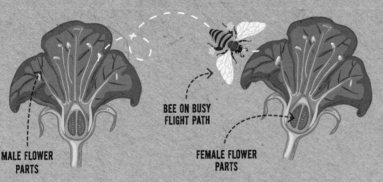

MALE FLOWER PARTS

BEE ON BUSY FLIGHT PATH

FEMALE FLOWER PARTS

3 As bees travel from flower to flower, pollen sticks to their bodies and is taken from one plant to another, where it sticks to the female parts. This is called cross-pollination.

1 Pollen is made by the male parts of plants called *anthers*.

2 The female parts of plants, called *pistils*, make the egg cells. When pollen meets an egg cell, they join together and create a seed.

4 Any animal that helps pollination is called a pollinator. As a reward, the pollinators get to take some of the pollen and nectar.

What is a bee?

Bees are a kind of insect, closely related to ants and wasps. Like all insects, they have a body that is split into three parts: head, thorax (middle part, with the legs) and abdomen (bottom part). They have six legs and two antennae – feelers that they use to smell and touch.

Insects are invertebrates, which means that they don't have a backbone. We humans use our backbone to help us move and hold us upright. Instead, insects have a special skeleton on the outside of their body. It's made from a hard material and is called an exoskeleton.

Bees have large compound eyes made up of hundreds of single eyes. They also have three 'simple eyes', which can tell how much light there is.

CAN YOU SEE LIKE A BEE?

How well can you see compared to a bee? Look straight ahead and spread your arms wide – try to stretch them as far back as you can.

Can you still see them? If you were a bee, you could. How useful is this? Because bees can see all around them, they can easily find food or know if an animal is sneaking up on them.

I've got double your wing power!

All bees have four wings: a pair of front wings and a pair of rear wings. Flies have only two wings, so this is one way of telling them apart. A few bees look like wasps. It can be confusing because sometimes you might see wasps drinking nectar from flowers too. Bees tend to be hairier than wasps and have special hairs to carry pollen.

Stingers

Only female bees have stingers. Once, stingers were long tubes used for laying eggs, but they evolved over time. Eventually, they became sharp, with a venom sack so that bees could protect their nests and honey from large predators wanting a sweet snack. Some solitary bees have such tiny stingers that they can't even get through your skin. Honeybees can sting, but they only do so if they feel threatened. In fact, a honeybee will die if it stings you.

HOW BEES GROW

Humans grow in size a little bit at a time, but bees don't – they go through different stages. This is called metamorphosis. From an egg hatches a larva, which has no wings. After eating lots and lots of food, the larva spins itself a cocoon. This is the pupa stage. In its cocoon, the eyes, legs and wings of the bee develop. When it is finally ready to come out, the adult bee chews its way out of the cocoon.

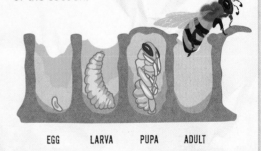

EGG   LARVA   PUPA   ADULT

Where bees live

Bumblebees, honeybees and some stingless and sweat bees live in large groups in nests or hives. Other bees are called solitary bees, and each female makes her own nest where she lays her eggs and leaves food for the hatching larvae. As we look at different types of bees in this book, we will also see where they live.

Types of bees

With so many types of bees, it can be difficult to tell them apart. Here are some of the most common types of bees that you might see and how to tell the difference between them.

Honeybees

Although there are only seven species of honeybee, these are the bees people know most about. The most common is the Western honeybee, which originally came from Asia or Africa, but has been moved by humans to every continent except Antarctica. Honeybees are shades of brown or black mixed with yellow and have stripy abdomens. They have a special basket on their legs to carry pollen.

Honeybees live in nests called hives. They live all year round, taking the nectar from plants in spring and summer and turning it into honey to eat over the winter. They store it in the middle of the hive, in hexagons made from wax called honeycombs. These are also home to the larvae.

These bees can be important pollinators, especially in areas where there are few wild bees. But some scientists are worried: because there are so many honeybees, they could be spreading diseases to wild bees, as well as eating a lot of the wild bees' food.

HONEYBEE

FACT
The average honeybee will make one-twelfth of a teaspoon of honey in her lifetime.

Inside a honeybee hive

POLLEN CELLS

THE QUEEN is the largest bee in the hive. She can lay up to 2,000 eggs a day.

THE QUEEN'S CUP becomes a queen cell when an egg is laid in it and the workers turn it into a bigger, peanut-shaped cell.

THE WORKERS are all the daughters of the queen. They do everything from carrying water back to the hive, to looking after the queen and her eggs, and guarding the hive from attackers.

THE DRONES are the males. They don't do any work and sometimes beg for food from the bees that are looking after the larvae.

HONEY CELLS

THE HONEYCOMB is made of six-sided hexagon shapes. This is a strong structure, and all the cells fit together perfectly without any gaps.

DRONE CELLS are bigger than normal workers' cells and can usually be found at the bottom of the hive.

LARVAE at different growth stages

BUMBLEBEE

Bumblebees

There are 250 species of bumblebees in the world. Most of them live in the Northern Hemisphere, and some in South America. Bumblebees have large, round bodies covered in soft hair. They can have stripes of yellow, brown, black, white or red. The bright colours are to warn other animals that bumblebees can sting.

Bumblebees live in a nest with a queen and between 50 and 400 workers. In cold areas, the workers die before the winter but the queen finds somewhere warm to hibernate (go into a deep sleep) over the winter. When she wakes up in spring, she finds somewhere to make a nest and starts laying eggs.

Bumblebees have long, hairy tongues to get nectar out of plants. They also have pollen baskets but they don't pack the pollen in as tightly as honeybees do. Because bumblebees don't need food over the winter, they make only a small amount of honey.

FACT
Some bumblebees fly between flowers with their tongue sticking out!

LATIN NAMES

We sometimes call animals and plants by different names in different countries. The buff-tailed bumblebee in the UK is called the large earth bumblebee in the USA. Scientists give all species a Latin name so that they know which living thing they're talking about. The Latin name for this bee is *Bombus terrestris*.

WHAT'S INSIDE A BUMBLEBEE NEST?

A bumblebee nest looks very disorganized compared to a honeybee hive, but this messy nest works for bumblebees. The queen builds the nest when she wakes from hibernation. She builds a few cells from wax and lays her eggs in them.

The workers are female and will be the first to help the queen grow the colony. They take any dead or dying bees outside and leave them by the entrance to the nest. This is to help prevent disease.

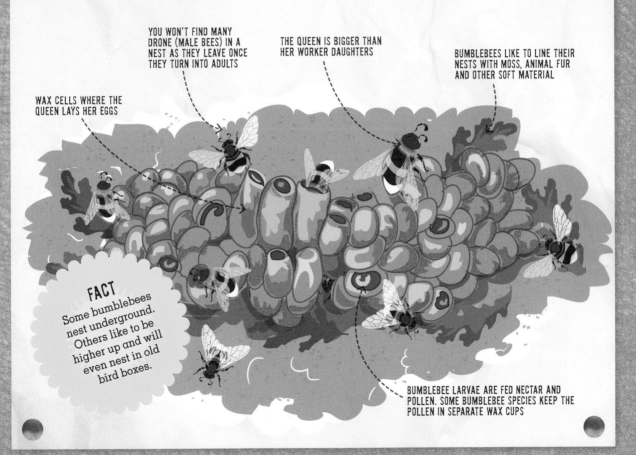

YOU WON'T FIND MANY DRONE (MALE BEES) IN A NEST AS THEY LEAVE ONCE THEY TURN INTO ADULTS

THE QUEEN IS BIGGER THAN HER WORKER DAUGHTERS

BUMBLEBEES LIKE TO LINE THEIR NESTS WITH MOSS, ANIMAL FUR AND OTHER SOFT MATERIAL

WAX CELLS WHERE THE QUEEN LAYS HER EGGS

FACT
Some bumblebees nest underground. Others like to be higher up and will even nest in old bird boxes.

BUMBLEBEE LARVAE ARE FED NECTAR AND POLLEN. SOME BUMBLEBEE SPECIES KEEP THE POLLEN IN SEPARATE WAX CUPS

Stingless bees

Stingless bees live in warm, tropical countries (around the middle of the Earth), in Australia, Asia, Africa and America. They live in nests and have a queen. Just like honeybees and bumblebees, the workers are all female and look after the nest. In some areas, such as Australia, farmers keep stingless bees for their honey.

FACT
Stingless bees do have a stinger, but it is too small to use for defence. But watch out – they will bite with their jaws if they are disturbed.

STINGLESS BEE

Who wouldn't want a tunnel in their garden chair?

FACT
Sometimes, carpenter bees can be unwanted visitors in people's homes. They can tunnel into unpainted doors or garden furniture to make their nests.

CARPENTER BEE

Carpenter bees

Carpenter bees are often confused with bumblebees because of their large size, but these bees tend to have a shiny bottom and a lot less hair. Carpenter bees don't live in a nest like honeybees or bumblebees, but lay their eggs in tunnels. They got their name because the females make tunnels by chewing through wood with their powerful jaws. Digging holes is hard work so carpenter bees sometimes try to nest in old abandoned tunnels that other bees have dug. Carpenter bees like to have nests close to each other. Sometimes, a mother bee will even share a tunnel with her daughter and they share the work of finding food and guarding the nest.

Leafcutters, mason & carder bees

Leafcutters, mason and carder bees all look very different but they are closely related. They carry pollen on bushy hairs on the underneath of their abdomen. When they've collected lots of pollen, they look like they've been dipped in bright yellow powder paint. It's fun to look out for them!

CARDER BEE

INSIDE A SOLITARY BEE BURROW

In this nest burrow of a leafcutter bee you can see that the bee has used little bits of leaves to make individual cells for each egg she lays.

The male eggs hatch out before the female ones. To stop them pushing into each other, the mother bee lays them in the right order, with the males at the front of the burrow and the females at the back.

All solitary bees search for straight tunnels where they can lay their eggs. The bees are named after the different things they use for lining the tunnel. Leafcutters nibble out little circles of leaves, mason bees use mud, and carder bees find animal hairs or fibres. (Carder is an old-fashioned word for someone who prepared wool for spinning.)

After lining the tunnel, the bees lay an egg at the back with some pollen and nectar next to it. Then they build a little wall from their lining material and lay another egg. They keep going until the tunnel is full. The eggs stay there over winter, and then the larvae hatch out, eat the food and dig their way out of the tunnel. If you make a solitary-bee hotel (see pages 30–31), these are the bees that will mostly likely come to stay.

Cuckoo bees

The cuckoo is a bird that lays its eggs in another bird's nest. In just the same way, the cuckoo bee lays its eggs in the nests of other bees.

You can tell a cuckoo bee because she doesn't have any way to collect pollen. Well, she doesn't have to! Instead, she sneaks into the nests of other solitary bees while they're searching for food and lays her eggs. When her larvae hatch, they eat the food of the host bee and probably the larvae too.

Mining bees

Mining bees are fairly small bees that like to nest in the ground. They make tiny burrows in soft sand where they can lay their eggs. If you see little mounds in the lawn with a hole in the middle, there is probably a mining bee in there.

Sit and watch to see if you can see the bee coming back to its hole with food. Instead of pollen baskets mining bees have long hairs on their legs, so they look like they're wearing yellow trousers.

Depending on the species of mining bee, they tend to prefer one type of flower. This is important to know for protecting bees. If that flower dies out, then the mining bee might too.

FACT
Some types of bumblebees are also 'cuckoos' that creep into other nests and lay their own eggs.

CUCKOO BEE

MINING BEE

Wake up, here comes breakfast!

Sweat bees

Found all over the world, sweat bees come in beautiful colours. Most of them are dark, but others come in bright shades such as metallic blue or green. Some are black with yellow faces while others have red abdomens and look like very small wasps.

Some sweat bees live in nests and have a queen. But most live quite happily as solitary bees, normally nesting underground. They line their tunnels with a waxy liquid that they squirt out to make their home waterproof.

Plasterer bees

Plasterer bees also like to keep their nests waterproof. They spit out a plastic-like liquid and smooth it all along the walls.

It can be difficult to identify a plasterer bee as some look like wasps. They can be very small, and come in yellow, black, white or metallic colours. Also, some don't have hairs to carry pollen. Instead, they swallow pollen and bring it back up when they get back to their nest.

SWEAT BEE

PLASTERER BEE

FACT

Why are they called sweat bees? They are attracted to humans and like to lick the sweat off our skin!

NIGHT BEES

Most bees are active only when the sun is out, but some plasterer bees go out at dusk or dawn. They have much bigger simple eyes than other bees, to help them see in dim light.

Bees & us

Did you know that bees have been around for about 100 million years? They lived during the time of the dinosaurs and might have been buzzing around the heads of iguanodons or other giant beasts.

Humans have known that bees are extraordinary creatures since the Stone Age, thousands of years ago. There are some cave paintings of bees in Spain that are 15,000 years old. In Turkey, scientists have found pottery with beeswax in it from nearly 9,000 years ago. Around 5,000 years ago, Ancient Egyptians used to put beehives on boats and row them down the River Nile to pollinate their crops. So bees were important to people back then.

Why do we need bees?

Bees help us to produce lots of delicious foods. Once a bee has pollinated a flower, the plant can grow seeds. Many plants make seeds inside a fruit. That's how we get amazing fruits such as apples, plums, strawberries and pears.

We think of peppers and cucumbers as vegetables not fruits, but if you cut them up, you can see the seeds. When their plants had flowers, bees visited them, then the fruit grew. With some vegetables like carrots and potatoes, we eat the fleshy roots instead of the fruit. These plants don't need to be pollinated by bees to grow the vegetables but do if you want the plant to make seeds.

What happens if there are no bees?

Vanilla was discovered in Mexico and Central America and it was grown by the people who lived there. Today it is grown in other tropical countries such as Madagascar and Indonesia. People thought the vanilla plant was pollinated by bees, but nobody knew which ones! Now the farmers have to pollinate the plants by hand. They bring the pollen to the egg cells, which takes a long time. This makes vanilla one of the most expensive spices in the world.

SHAKE
SHAKE
SHAKE
SHAKE

Buzz pollination

Bumblebees are important for pollination because they can 'buzz-pollinate'. In some fruit and vegetables, such as blueberries, kiwifruit and tomatoes, the pollen is stuck onto the anthers (the male parts of the plant). If the bee wants the pollen, it has to hold on tightly and vibrate its wing muscles 100 times a second to make it drop off. That's fast! Only some bees are able to do this.

Make a FLOWER COLLAGE

If you see a lot of bees buzzing around wildflowers in your garden or a friend's garden, ask a parent or carer if you can pick a few of them. Carefully place each flower between two pieces of paper and weigh them down with a heavy book. After a few weeks, when they are dry and flat, you can make a pressed-flower collage. You'll have a beautiful reminder of some bee-friendly flowers.

Why are bees struggling?

There aren't as many bees in the world as there used to be. In some parts of the USA, half of all the wild bees have disappeared. This is worrying, because fewer bees means less pollination – and fewer fruits and vegetables.

The biggest problem for bees is that there are not as many plants around. In the past, much of the land was left wild and covered in flowers. Nowadays, we grow lots of food crops on the land. Bees love food crops – but the problem is how we grow those plants.

Sometimes farmers grow the same crop over a huge area. This is called industrial agriculture. The plants only flower for a couple of weeks a year, which means that any bees there will quickly starve.

In industrial agriculture, farmers use many pesticides on their crops. Pesticides are chemicals that kill pests that like to munch on the plants, such as slugs and aphids (tiny insects such as greenfly). But these pesticides don't just target the pests, they also kill helpful insects, including bees.

Oh no, not wheat again!

FOR SALE

Honeybee hazards

Honeybees have struggled for survival in the USA. Beekeepers have reported entire hives being abandoned. All the bees have disappeared from them. Scientists call this weird happening 'colony collapse disorder'. They have suggested many ideas why this occurs but nobody knows the answer yet.

Honeybees also catch diseases easily, and these can spread quickly. The *varroa destructor* mite is at the top of the nasty list. This parasite attaches itself to a bee's body and sucks its blood – yuk! If a honeybee has this mite, it is more likely to become infected with a disease.

Wild bees

Wild bees are even more threatened than honeybees. Not only are they suffering because of too many pesticides and the effects of industrial farming, but our changing weather could also make it harder for them to survive in the future.

Hey! You've just taken my dinner!

Honeybees v wild bees

In places where there are large numbers of honeybees, it can cause problems for wild bees. If beekeepers move their hives into an area that has just a few flowering plants, the honeybees quickly grab a lot of the food for themselves and don't leave enough to go around.

PESTICIDE PROBLEMS

For several decades, so many pesticides were used in Szechuan, China, that now there aren't enough wild bees to pollinate the orchards of apple and pear trees. Now, the farmers have to pollinate the trees by hand. They carry a pot of pollen and little paintbrushes up into the trees and brush each flower with the pollen.

Buying bees

If farmers have lots of crops that need pollinating, but there aren't enough bees around to do it, they can order in some help. In the USA, beekeepers travel around with their honeybee hives and place them in the middle of the crops to pollinate the flowers.

In California, there are huge areas of land with almond trees but there aren't enough bees. Every year, nearly two million hives are put on the back of trucks and driven in. The bees stay there for a few weeks and then are moved to a different crop. Scientists believe this is one of the threats facing honeybees – all that travelling could be making them stressed.

Hard-working mason bees

If we had to pick the best pollinator, the mason bee would probably be top of the list. Mason bees carry pollen loosely on their bodies so it easily falls off onto the next flower, helping with pollination. They also visit more flowers than honeybees and work even when it's cold (honeybees tend to be weather wimps).

If you want a super pollinator call Mason Bee!

More and more farmers are buying mason bees to help pollinate their crops. They buy them as pupae in nest boxes and keep them in the fridge until it's the right time to drive them to the fields for their pollination work.

Newbies cause disaster

Bumblebees have also been sold in boxes and moved to places where they can help with pollination. But sometimes this can cause huge problems for the local wildlife. In the mid-1990s, buff-tailed bumblebees were taken to Chile in South America and released in greenhouses to help with pollination. They escaped and spread diseases to the bumblebees that had always lived in the country.

How to become A BEE SCIENTIST

A scientist who studies insects is called an entomologist. I learnt how to become an entomologist by going out and searching for different bugs. Observing insects and their behaviour is a great way to learn more about them.

For example, in 2010, a group of British schoolchildren aged between 8 and 10 discovered that some bumblebees use colour and location to learn which colour flowers give lots of food.

You can use the pull-out observation sheet opposite to keep a record of the bees you see and the flowers they like to visit.

FACT
Bumblebees can fly up to 20km (12½ miles) away from their nest to find food – they get around pretty quickly!

Predators of bees

It's not just humans that love honey, lots of animals do too. Bees have to be able to protect their nests, which is one of the reasons why their stings are so painful. Some creatures are even bee predators – they feed on the bees themselves.

Honey bears

Even with the risk of stings, some animals just can't resist honey. The Malayan sun bear is known as the 'honey bear'. It uses its long claws to dig into tree trunks and its 20-cm-long (nearly 8-in) tongue to reach the sweetness inside. Bears also eat bees and the larvae because they're a great source of protein.

SUN BEAR

Insect bee-predators

Wasps and hornets are natural predators of bees. Beewolves (also called bee-hunters) are a group of wasps that collect bees to feed their young. The female stings the bee with venom that paralyses it (stops it moving), and carries it to her tunnel nest. She leaves the bee with an egg. When the egg hatches, the wasp larva will eat the bee alive.

Bee-eating birds

Some birds are very good at eating bees without getting stung. Bee-eaters are found in Africa, Asia, Australia and Europe. These beautiful, brightly coloured birds catch their prey mid-air. They hit the bee over and over again on a branch until the stinger falls off and most of the venom leaks out, letting the bird eat it in safety.

The greater honeyguide is an African bird that has some strange food tastes. Rather than eating honey or adult bees, it feeds on the bee larvae in a honeybee hive. It even eats the beeswax. The honeyguide is one of the few birds that can eat wax without becoming ill.

Hey everyone, there's a hive over here!

This clever bird has a neat trick to get into the hive. In Mozambique, it shows the Yao hunters where the hives are by chirping and flashing its wings. Because the local honey-hunting humans know the bird, they chirp back, and the bird takes off to guide them to the hive. Once they find it together, the hunters smash open the hive to take the honey, and leave all the other parts for the honeyguide. It's a great bit of teamwork.

HOLDING OFF THE HORNETS

Hornets (large wasps) are predators of honeybees, but some honeybees have come up with ways to fight back. Japanese honeybees surround a giant hornet and vibrate their muscles so quickly that the hornet 'cooks'. Another type of honeybee (from Cyprus) forms a tight ball around an attacking hornet to stop it breathing.

Why bees love flowers

Bees and flowers have lived around each other for millions of years. Plants have developed cunning ways of getting bees to visit them – by producing nectar and making their flowers stand out.

Flower shapes

Flowers come in a wide variety of shapes, and while some attract lots of different insects, others attract only a few, or even just one type.

PEA PLANT WITH PEA FLOWERS

BOWL-SHAPED (LIKE POPPIES)

POPPIES

These flowers have a large, open ring of anthers in the centre with lots of pollen. Many insects love this type of flower because it's very easy for them to get to the pollen. Bees run around the inside of the flower in a circle to collect it all up.

LIP-SHAPED (LIKE PEAS) These flowers have a lip at the front that insects use as a landing platform. To get to the nectar, the bee has to push past the pollen-covered anthers, which transfer pollen to the back of the bee's neck. It is difficult for the bee to get the pollen off its body, so the pollen is carried to the next flower and pollinates it. It's a clever trick!

BELL-SHAPED (LIKE BLUEBELLS) Flowers like this have long tubes with nectar at the very end. They are a favourite of bees with long tongues, such as some of the bumblebees, mining, leafcutter and carpenter bees.

DAISY-SHAPED The centre of a daisy is made up of many tiny flowers. Because they are tightly packed together, larger bee species such as bumblebees can find it quite difficult to get to the nectar and pollen. But these flowers are perfect for smaller species, such as mason bees.

FLOWER COLOURS & PATTERNS

The colour of flowers is also very important. With their compound eyes made up of hundreds of lenses, colours look different to a bee than they do to us humans. Bees can also see ultraviolet light, which we can't.

Under ultraviolet light, some flowers have patterns that are invisible to us. Luckily, there are cameras that can photograph them, allowing us to see what a bee sees. Under ultraviolet light, a dandelion has a different-coloured centre that shows bees exactly where to land.

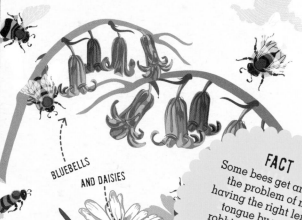

BLUEBELLS AND DAISIES

FACT
Some bees get around the problem of not having the right length tongue by nectar-robbing. They chew a little hole at the base of the petals and get to the food that way!

NORMAL LIGHT

ULTRAVIOLET LIGHT

Attractive scents & communication

We love the scent of flowers, but for bees it's pretty important. Bees prefer sweet-smelling flowers, and their scents can lure in pollinators over long distances.

Plant smells are at their strongest during the day, when bees are out, and at the time of year when the flowers are ready to be pollinated. Bees can smell some flowers that are more than a kilometre (more than half a mile) away.

Flower mimics

Bee orchids are sneaky flowers that trick bees into coming to visit them. Each flower looks and smells like a female bee. When an excited male bee comes along, he tries to mate with it, getting covered in pollen. The unsuccessful male then flies to the next flower and pollinates it.

Buddleja

We love to introduce exciting plants from other countries into our gardens. But sometimes, unexpected things happen. One example of this is the buddleja, which was originally from China. This beautiful bush is also known as the butterfly tree because butterflies love it – and so do bees, as they are attracted by its strong scent. But although it is great for pollinators, it spreads like wildfire and in some countries, including the UK and the USA, gardeners are discouraged from growing it to stop it spreading too much.

BUDDLEJA SPREADS EASILY

Pheromones

Bees also use scent for communication – after all, they don't have voices! Their bodies make chemicals called pheromones, which are released when the bees need to communicate with each other. Social bees which have a queen use pheromones to control what the rest of the bees are doing.

The waggle dance

Honeybees have a special way of telling others where to find food – they do the waggle dance. Worker bees really know how to shake it. The basic move is a figure of eight with a wiggle on the middle line. The direction of the wiggle lets the others know in which direction to go, and how long she wiggles tells them how far to go.

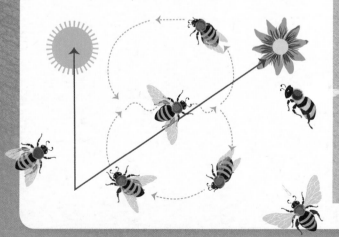

How to volunteer FOR BEES

One of the best ways of learning more about bees is by volunteering with a nature group. Many organizations around the world do great work for wild bees, and you can help them. Are there any pollination charities in your country? You could talk to your local beekeepers' association and find out about the work they're doing to protect nature.

Perhaps there are local groups that go out and do bee surveys. Many organizations rely on people like you to help with science surveys. By volunteering, you can find out more about bees and meet interesting people who can teach you new skills. A lot of people who work with wildlife started out by volunteering.

You could also become a campaigner and speak up for the bees. Write letters to politicians to tell them why bees are important and ask them to help protect our pollinators.

FACT
Scientists have found that diesel fumes from cars can make flowers smell different to bees so they don't realize the flowers have food.

Make a solitary-bee hotel

You may not have space for a beehive, but there are plenty of things you can do to encourage bees into your garden or backyard. Bee hotels are very simple to make, and great fun too. You can get really creative with this project. Don't worry too much about making it perfect – bees aren't fussy!

WHAT YOU WILL NEED

- Old bamboo gardening canes about 9 cm (3½ in) long (don't worry if the soft inner part is still there – the bees will be able to burrow their way in)

- Dead hollow plant stems (such as reeds from a riverbank, or elder stems). Any width between 2 mm (¹⁄₁₆ in) and 1 cm (³⁄₈ in) is perfect, because bees come in all shapes and sizes

- String

1 Using the string, tie the canes and hollow stems into a tight bundle.

This will look grand when it's finished!

TIP
Don't forget to move your bee hotel to a cool, dry place in a shed or garage over winter. You can move it back outside again once the risk of frost has passed.

2 Place your bee hotel in a sunny position. It should be at least 1 metre (3⅓ ft) above the ground so that it stays warm and dry.

3 Your bee hotel is now open and ready for guests!

BEE & BEE

It's the bee's knees!

MAKE A BEE MANSION

If you have more time, you can turn your bee hotel into a snazzy bee mansion. Ask an adult to cut out five pieces of wood 9 cm x 9 cm (3½ in x 3½ in). Make a square box frame using four of the pieces for the sides and one to make a back panel to keep the bamboo canes dry. Ask an adult to use nails to fix it together. When it is finished, you can fill it with bamboo canes as for the bee hotel. Remember that it does not have to be neat and tidy, because the bees really don't mind. Fix a bracket to the back to hang it securely to a wall or post, and watch for bees coming to check out their lovely new home.

Helping bees to nest

You can help solitary bees by growing special plants and putting out useful materials they can use to make their nests. Helping the bees can be rewarding, especially when you see them flying back and forth.

This garden's great for building materials.

Natural nests

Don't worry if you don't have the materials for a solitary bee hotel (see pages 30–31). With an adult's help, cut down the stems from plants such as roses, bramble and teasel in your garden. If you leave them somewhere sunny, bees will burrow into the middle of the stems and nest there.

You can also leave out a little platter with mud for mason bees – they will use it to line their burrows. Some plants provide nesting materials. Roses, azaleas and redbud are great to grow. Leafcutter bees like munching holes in the thin leaves, which they use to make their nests (see page 15). Lamb's ear is a plant with soft fluffy leaves. Carder bees like to use the hairy fibres to line their tunnels.

Nesting patches

Some bees, such as mining bees, nest in the ground, so being able to reach the soil is important. Clear away the plants from little patches of ground, from about 5 cm (2 in) to 30 cm (nearly 12 in) wide. Make sure your patches are in a warm, sunny spot.

Start composting

Make a compost heap! Bumblebees love to nest in warm, dark places and, although a lot of them end up underground in mouse burrows, a few can get confused and find their way into your compost heap. If you find a bumblebee nest in your heap, remember that the nest will die in August or September.

BECOME A BEEKEEPER

If you want to learn more about honeybees, why not become a beekeeper for a day? There are many places that run beekeeping courses where you can learn how to care for a hive. Some primary schools have their own beehives, which are looked after by both the teachers and the students. It's a great way of learning from real life. Could you ask your school to do the same? Make sure that you also grow plenty of different plants to support your hives and the bees.

Make a bumblebee nest from a plant pot

It can be difficult to make a nest for bumblebees as they like to nest in different places. Some species nest underground in old mouse holes, or in the base of tussocks of grass. Others like to nest above ground and can be found in old bird boxes or trees. If you want to try to build a bumblebee nest, this is the best design to use.

WHAT YOU WILL NEED

- Plant pot wider than 20 cm (8 in) with a hole in the bottom

- Piece of heavy slate or tile to fit on top of the pot

- Piece of rubber tubing about 2 cm (¾ in) in diameter and 40 cm (15¾ in) long

- Soft moss or small animal bedding

- Scissors or bradawl

- Spade

1 Ask an adult to help you to make six holes in a row on the underside of the rubber tubing to let rainwater drain away.

2 Find a shady spot near a hedge or a flower bed and bury the tube in a small trench about 5 cm (2 in) under the ground but with both ends sticking out by about 2 cm (¾ in).

3 Place the bedding above one of the tube ends and put your plant pot over it, making sure that you cover all the bedding and the end of the tube. Put the tile on top. This is the main part of the nest. The other end of the tube is the entrance.

A visit from the queen

Bumblebee queens hibernate over the winter in underground burrows. When they wake up, their first job is to find a new nest site. You can see them flying low across the ground in zigzag patterns and stopping to investigate every hole and crevice. Who knows? Maybe a queen will stop to check out your home.

Even if bumblebees don't move in straight away, leave your nest for a few years. A mouse might use it. Bumblebees seem to like the smell of old mouse burrows and might move in afterwards.

Growing bee-friendly plants

Because industrial farming has made things worse for bees, gardens have become more important. We need to make sure that the flowers we grow are bee-friendly. Lots of sweet-smelling herbs, fruit and vegetables are great for bees too.

The best plants to grow are flowers with simple petals. Fancy blossoms often have double petals, and it is hard for bees to reach the pollen inside. Try to grow plants that flower over different months. Bulbs that come out in the early spring will help early bumblebees, while late-flowering shrubs such as ivy or goldenrod give nectar and pollen in late summer and autumn.

Tasty plants

You can help the bees and make your dinners delicious at the same time. Bees go crazy for herbs. Look out for different kinds of mint: spearmint, peppermint, apple mint and lemon mint all have green spikes of little flowers to draw in crowds of bees. Try growing them in pots instead of in the ground, as their roots spread quickly.

FACT
A group of pesticides called neonicotinoids are really bad for bees. Check with your garden centre that they haven't been used on any plants that you buy.

Aah, the sweet smell of success!

Sage, wild thyme, oregano, chives, lavender and comfrey are all great plants for bees and can be grown in small pots. Rosemary turns into quite a big bush so it is not good for small spaces.

If you grow your own fruit and vegetables, you'll be helping the bees as well. You could grow tomatoes in a grow-bag, a row of salad crops or an apple or pear tree. This way, you can look after yourself – and the bees.

HELPING BEES AT SCHOOL

If you have large school grounds, why not ask if you could plant some bee-friendly flowers? You could grow some plants from seed in the classroom to learn about plant science (botany), and then plant out the seedlings when they've grown big enough.

SEED BOMBS

Seed bombs are great fun. They are little balls of clay or soil filled with seeds. Seed bombs are a clever way of putting life back into the ground. The soil or clay gives a helping hand to flowers that may not grow easily in poor ground soil. Always ask permission from the person who owns the land first.

1 To make your seed bomb, mix three handfuls of clay with a handful of seeds and five handfuls of compost. Wildflower seeds are the best to use for this.

2 Mix in water slowly so it feels hard enough to make into little balls, and not too runny. You don't want to make seed soup!

3 Leave to dry in the sunshine for a few hours. Then throw down your seed bombs.

PACKED WITH SEEDS

Wonderful wildflowers

Wildflowers are brilliant for bees. They're easy to grow and can survive cold weather. And, of course, they look beautiful.

Insects pollinate eight out of every ten wildflowers. Some rare bees only take nectar or pollen from one type of flower and would starve if the plant disappeared. Wildflowers also provide nesting sites. If left over winter, they can be a safe place for other insects, such as ladybirds, to hibernate.

Growing wildflower strips is helpful for bees. Flying takes up a lot of energy, so if they can have a snack on the way, it's easier for them to get to new places. You can make strips of any size, along fields, paths or in a garden.

FACT

Put the mower away. Cutting your lawn only twice a year can let wildflowers naturally take over your lawn – much prettier than just green grass.

Phew, a snack stop at last!

FERTILIZER FACT

In countries with a lot of industrial farming, there are far fewer wildflowers than there were in the past. Wildflowers used to grow on farmland but struggled to survive when chemical fertilizers were introduced. The fertilizers add nutrients to the soil that help grasses grow instead of the flowers.

MAKE A WILDFLOWER WINDOW BOX

Luckily, you don't need a huge amount of space to make your own wildflower strip. Simply find a planter with some holes in the bottom and fill it with compost. Make a trench down the middle and scatter your wildflower seeds. (Remember to choose seeds for flowers that grow well in your area.)

Gently cover the seeds with soil, pat down and water. Put your window box outdoors in a place that gets plenty of sunshine, and watch those flowers grow.

SCATTER A MIX OF WILDFLOWER SEEDS

Delicious dandelions

Love those weeds! Some people say that a weed is a flower growing in the wrong place, and this is a great way to think about weeds. Dandelions are probably the best example. Many gardeners think that dandelions make their beautiful lawns look ugly. But lawns full of grass alone don't do a lot for wildlife.

This is why I say 'put back the posies' – well, a few flowers anyway. You can start with dandelions. They bloom early in spring, just when bees are starting to wake up. Dandelions give the bees an easy meal of nectar. So please spread the word about these helpful weeds.

Living walls

You don't need to have a lot of space in your garden to provide the perfect buffet for bees. How about going upwards? Living green walls covered in plants don't just look amazing and attract pollinators, they can also do some other helpful things.

Cities and towns are usually warmer than the countryside because we build with materials such as concrete, which hold on to heat. Scientists found that the surface of a green wall can be 10°C (18°F) cooler than a bare wall. So living walls help keep cities cool in the summer.

Living walls also add a layer of insulation (protection from cold) that keeps the house warmer in the winter. They protect the walls from heavy rain, give clean air and even block outside noise.

You can make a simple green wall by putting up some trellis and growing lovely creeping or climbing plants such as honeysuckle, wisteria or jasmine. You could also try growing giant sunflowers.

What are you hanging about for?

Hedges

Growing a hedge instead of a fence will not only help make your garden private but will also be brilliant for wildlife. You can include small trees but make sure that they don't get overgrown and take over the garden. Hawthorns, wild cherry, crab apple and wild roses attract pollinators but also provide food and shelter for other animals in the winter. Pussy willow and blackthorn are some of the first to flower in the spring, which is great for the early bumblebees. Having a mix of different shrubs makes your hedge look wonderful and provides lots of different types of pollen that bees can feast on.

Hanging baskets and pots are a brilliant way of growing plants on a balcony or window ledge. Make sure you use pots with holes in the bottom so that water can drain through, and remember to water them regularly in dry weather.

NO NEED TO TIDY UP

FACT
A messy garden is a wildlife-friendly garden. Little piles of leaves, some long grass and a pile of logs will encourage all kinds of creatures to move in.

Create a wildlife area

You can help bees by providing them with water to drink, avoiding pesticides and creating wildlife areas.

Some people use pesticides in their gardens so that their flowers look beautiful all year round, but if you want to help bees, then put away those sprays! If you have a lot of pests, try washing down the plants with diluted washing-up liquid and water.

FUKUOKA HALL

Nice place – do you come here often?

MAKE A WATER BATH

All animals need to drink water, including bees. If you don't have enough space in your garden for a pond, you can make a water bath. Find a shallow dish, a plant pot and some stones. Turn the plant pot upside down and place the dish on top of it. Fill it with stones so that if bees get stuck they can climb onto them and will not drown. Add water, and enjoy seeing your thirsty guests come for a drink.

Green roofs

Green roofs have become more popular in recent years. Chicago City Hall, USA, has 20,000 plants on its roof. The ACROS Fukuoka Hall in Fukuoka City, Japan, has 100,000 square metres (nearly 25 acres) of greenery on its terraced roofs.

These roofs are beautiful and helpful too. The plants suck up pollution, improving the quality of the air, and they keep buildings cool in the summer. The laws in some cities, including San Francisco, USA; Cordoba, Argentina; and Toronto, Canada, mean that any new buildings have to include green roofs.

Community meadows

If you haven't got a garden or much outdoor space at home, or fancy a bigger project, how about creating space for wildlife in your community?

While our gardens might be full of flowers, our parks might not be so lucky. Most people want short grass where they can play games instead of long grass and flowers. But there can be room for untidy areas for wildlife as well as tidy areas for play.

LOCAL MEADOW

Talk to your local park managers and ask if they would be interested in making a bee-friendly wildlife meadow. Can you encourage them to work with a local wildlife charity to create some new amazing spaces for bees?

More and more people are becoming aware of how important bees are, and they want to help. Some cities are banning the use of pesticides, and people are creating rooftop gardens to brighten up grey, concrete cities. These are just some simple ideas and tips to get you started on your journey to be a bee hero!

GOOD FOR BEES AND PEOPLE

Honey & other bee products

This nectar's yummy, you've got to try it.

Whether you spread it on your toast, mix it in your porridge or use it to sweeten your tea, there are lots of reasons to love honey. But bees make lots of other useful things. You might have heard about beeswax, but what about propolis, mead and bee venom?

How is honey made?

Honeybees swallow nectar and store it in a special stomach. When they get back to the hive, they bring up the honey into another bee's mouth, where it will get passed around to other bees. Much of the liquid is removed, so it becomes sticky. Then the honey is stored in cells in the hive. If it's still not sticky enough, honeybees fan their wings over the cells to dry it out before covering it with wax.

HONEY FOR MUMMY

Archaeologists found 3,000-year-old honey in an Egyptian tomb that could still be eaten! A chemical called hydrogen peroxide can develop in honey and stop it from going bad. Some scientists use this chemical to help rockets to take off. But don't worry, honey is perfectly safe to eat.

BEST BEFORE 1,000 BCE

STRANGE BLUE HONEY

Some honeybees in France made the news when they started making bluish-green honey. It turned out they were feeding around a waste plant that was getting rid of coloured chocolate sweets. The colours from the candied coating were turning the honey a funny colour – much to the annoyance of the beekeepers.

Beeswax

We use beeswax in lots of ways: for polishing furniture or shoes, making candles or as a coating for cheese to stop it from going bad. Long ago, it was even used to fill holes in teeth. Scientists found a 6,500-year-old cracked tooth filled with beeswax. Talk about trying to get out of seeing the dentist!

Propolis & mead

Propolis is made by bees when they mix their saliva with beeswax and tree resin to make a glue. Some violin makers use it to varnish their instruments.

Bees help us make an alcoholic drink called mead, made from fermented honey mixed with water and spices. The ancient Greeks called mead 'ambrosia', or the 'nectar of the gods', because they thought it came from the heavens. Scientists tested the insides of 9,000-year-old pottery jars from China. They found traces of a drink made from mead, grapes and rice, which makes it one of the world's oldest alcoholic drinks.

Could venom be useful?

A bee sting can be extremely painful – the venom the bee injects into your body is an acid. It can be dangerous if you're allergic to the venom. But scientists think it could be useful in medicine. With a bit more research, bee venom might eventually be used to treat diseases such as cancer.

BEE STING

Other pollinator pals

Although bees carry out a lot of pollination, they're not the only ones. There are many other pollinators, and some of them might surprise you. Did you know that we wouldn't have chocolate without flies? A tiny fly midge is the only pollinator of cacao, the bean that's used to make chocolate.

Butterflies, moths & wasps

Butterflies and moths have long tongues so they can pollinate flowers with long tubes. Many moths are night pollinators and are attracted to flowers that smell sweet during the night, such as jasmine or verbena. Some hawkmoths hover in mid-air as they sip nectar using their long tongues and may be mistaken for tiny hummingbirds.

Even wasps can be great pollinators. Although some hunting wasps eat other insects, they may visit the occasional flower for a quick burst of energy-rich nectar.

Fig wasps rely on figs to breed, and pollinate them in return. The female fig wasp makes her way into a fig before it ripens. As she pushes her way in, she pollinates the flowers on the inside of the fruit, lays her eggs and then dies. The eggs hatch, and the young wasps grow up inside the fig and mate. They dig a tunnel to get out, the males die, and the females fly off with the pollen from that fruit to another tree to start all over again. Don't worry – all the wasps leave the fig before you eat it.

HAWKMOTH

Beetles & flies

Beetles are some of the oldest pollinators – they were around millions of years ago when the very first flowering plants evolved. Even now, beetles are important pollinators of some of those ancient plants that are still around today.

Magnolias are a perfect example. They appeared before bees existed and so they had to attract beetles. Magnolias have large, open flowers, which beetles love. They even have tough petals in case the beetles accidentally get carried away and try to eat part of the plant with the pollen.

Flies also pollinate flowers. Some flowers try to be especially stinky to encourage flies to visit them. These aren't your regular flowery scents. The rafflesia, the world's biggest flower, gives off a smell like rotten meat to attract swarms of flies. The flies turn up expecting a tasty dead body but have to go away disappointed. The flower gets pollinated though!

FACT
Bats like large, pale flowers that open at night and smell strongly of fruit. Because bats are nocturnal (awake at night) this helps them find the flowers in the dark.

Birds & bats

It isn't just insects that visit flowers. Hummingbirds and bats are also pollinators; both use their incredibly long tongues to reach into flowers. A bat's tongue can be longer than its body and can reach into the flowers of fruits including banana, mango and guava. Monkeys and lizards have also been spotted pollinating plants.

No bees round here? Leave it to me!

Index

Picture credits

Alamy Stock Photo/Acro Images
6, AfriPics.com 46, Binu Balakrish
(top), Richard Becker 15, 17 (botto
Julian Brooks 16 (bottom), Peeras
Chaisanit 14 (left), Mircea Costin
John Glover 43, Lucy Pope 40, Th
Wildlife 14 (right), WILDLIFE GmbH

Getty Images/Auscape 24, Alista
23, Kevin Frayer 21.

Shutterstock/aarud 19 (backgrou
19 (top), Ian Grainger 12, David F
kosolovskyy 33, periphoto 28, pho
(bottom), Frank Reiser 45, Barbar
Storms 17 (top), yyama 42.

Kiddish font courtesy of Matt Bru

AUTHOR'S ACKNOWLEDGEMEN

Thank you to Glenn fo
being my rock, and to Josh
Maya for their inspiratio